MODERN EYE SURGERY

MODERN EYE SURGERY

M Clement Hall

MODERN EYE SURGERY

ISBN 978-1-105-90063-1

CONTENTS

EYE CATARACT SYMPTOMS

The origin of the word *cataract* is obscure, but "obscure" describes the condition. The earlier doctors thought a veil had formed between the front of the eye (*cornea*) and the lens, so the lens appeared clouded. We now know the change is in the lens itself.

The Lens

It is tempting to think of the lens in the eye as a simple inert object of uniform consistency, like a magnifying glass, however, biological systems are rarely simple. In fact the lens is a highly complex alive structure, composed of several materials, in several layers, and the great wonder is that light penetrates these layers.

Not a very attractive thought, but consider your eye is rather like an onion (also a biological wonder). The greater part of the lens is composed of layers (*laminae*) of cells, just as in the onion, these cells are long, thin, extend from the front to the back of the lens, and contain protein (*crystallin*) which assists the passage of light and probably also has a metabolic function.

These cells are contained by a thin *capsule* made of collagen fibers, the same material that's found in bone, tendons and cows' hooves, yet it's organized in such a manner light penetrates through it. The elastic capsule is attached at its margin to a muscle (*ciliary body*) which by contraction will alter the shape of the lens (*accommodation*) as in focussing.

A layer of cells on the front of the capsule do the work needed to preserve the health of the lens, which is the same as any other part of the body whose cells require fluid and chemicals to keep them alive and functioning – and the lens is very much "alive."

The Cataract

The essential feature of the cataract is a reduction in light passage or a distortion of the light's rays due to some change in the nature of the lens itself. This may be found at birth, it is associated with a number of medical conditions, noticeably diabetes, and with injury, but commonly it is one of the features of aging.

The change may occur in the center of the lens (nuclear cataract), on the outside of the lens (cortical cataract), or beneath the capsule (subcapsular cataract); all have the effect of diminishing acuity of vision and some reduce perception of blue color to the extent the patient is delighted with the color of the sky after their surgeon has replaced the defective lens.

EYE CATARACT SURGERY

The essential problem is obstruction of the passage of light. In times past, the cataract was treated by simple removal of the lens, this permitted unfocussed vision, an improvement on blindness, but sometimes also meant the whole interior of the eye spilled out; certain unlicenced practitioners in India were said to be very expert at doing this with the use of a sharp thorn.

Next came removal and replacement of the lens; not so long ago and there are doctors still in practice who remember the days when this was a major operation, with a large incision into the eyeball requiring stitches and for the patient to lie seven days in bed with his head fixed between sandbags in case the lens popped out. But things have improved since then, although there are still two stages – removal of the defective biological lens and replacement with an artificial lens. Whereas it used to be suggested the operation should be put off as long as possible, it is now generally recommended that it is performed when changing glasses no longer meets the patient's needs – glasses cannot compensate for the distortion caused by the lens of the eye.

Circumstances differ, as do patients and surgeons, but the procedure is now pretty much standardised in a process called phacoemulsification. The eye is "numbed" by local anesthetic drops, using an operating microscope the surgeon makes a tiny (less than 2.8 mms) incision, an ultrasound probe that vibrates 40,000 times a minute is inserted, the cataract is liquefied, sucked out from the intact capsule, into which an artificial lens is implanted. The wound is so small it heals without stitches.

The Implanted Lens

There have been tremendous improvements in the physical nature of the lens implanted, there are many persons in need of the procedure, and much industrial research is performed to meet the demands of the market.

Essentially, whichever lens is used, it is flexible and can be folded to be passed through the tiny incision to unfold when inside the lens capsule. The simplest pattern of lens (monofocal) is to provide clarity of distance vision, and reading glasses are required for close vision. The more complicated version, akin to spectacles, allows the eye to focus at differing distances (multifocal), others are "hinged" at the margins permitting a degree of muscular accommodation.

Distortion of the shape of the cornea (*astigmatism*) is common, it causes blurring of the image and reduces clarity of vision. To an extent this can be corrected by the lenses in glasses; in the implant it is corrected by a custom-shaped lens (*toric*) which will not eliminate the need for reading glasses.

LASEK and LASIK

Although there are differences between them, most persons know these procedures more simply as *laser eye surgery*. Their purpose is to reduce or eliminate the need for eyeglasses (spectacles) by reshaping the anterior portion of the eyeball (*cornea*) if it is the cause of a defect in vision, such as short sightedness (*myopia*), long sightedness (*hyperopia*), or distortion due to an asymmetry of the cornea (*astigmatism*).

The cornea

It is not generally understood that the cornea, at the front of the eye, is an essential component of the focusing mechanism. Shaped like a section of a sphere, it is only about a half millimetre in thickness, but is composed of three distinct layers. On the surface is a layer five cells thick (*epithelium*), which is capable of repair if injured. Beneath that is the main component of the cornea (*stroma*) composed of collagen fibers, as is a tendon, arranged in two or three hundred sheets, and formed and supported by cells (*keratocytes*). On the inner surface of the cornea is a one cell thick *endothelium* part of whose function is the control of fluid into the cornea, essential to preserve its transparency.

LASEK is the acronym for *Laser-Assisted Sub-Epithelial Keratectomy*, whereas LASIK is the acronym for *Laser-Assisted In situ Keratomileusis*.

In both of these procedures the cornea is surgically re-shaped by incisions made into it. The main difference between Lasik and Lasek techniques is the thickness of the flap which includes corneal stroma tissue in Lasik (100-180 microns) and only epithelial tissue in Lasek (50 microns); healing time is a few days faster with the Lasik procedure.

The Lasik method might be preferable for those with more corneal tissue and the Lasek method for those with less, with the potential for complications if an inappropriate choice is made.

Indications
Ages between 18 and 40, vision must be less than -14.00 diopters of nearsightedness, less than +6.00 diopters of far sightedness, and less than 6.00 diopters of astigmatism; stable prescription; no history of eye disease; corneas within a treatable shape range; adequate tearing in both eyes.

Complications
The Eye Surgery Education Council (ESEC) reports less than 1% of treated patients experience serious problems if proper screening is performed and an experienced surgeon conducts the procedure. Between 3 and 5% of patients experience less serious problems which are correctable. There have not been any reports of blindness resulting either from Lasik or Lasek surgery.

LenSx

In laser surgery, a laser beam vaporizes the water content of the tissue, and there is no loss of blood.

The LenSx Laser

From the manufacturer

The LenSx® Laser is a fully-integrated, image-guided femtosecond laser designed specifically for refractive cataract surgery. [femtosecond is the SI unit of time equal to 10^{-15} of a second, or one millionth of one billionth of a second.]

Using a customizable 3-D surgical platform, it allows you to visualize, customize and perform many of the most challenging steps of cataract surgery, such as: Anterior capsulotomy; Lens fragmentation; All corneal incisions.

The LenSx® Laser is the first femtosecond laser cleared by the FDA for use in cataract surgery. The system brings a new level of precision to these surgical steps through a number of high-tech features:

Real-time video imaging with integrated OCT. Provides three-dimensional visualization of the entire anterior segment during docking, planning and procedure.

Curved patient interface. Designed for patient comfort, ease of use and optimal laser performance.

Intuitive touch screen graphic user interface. Allows each step of the procedure to be easily planned, customized and executed.

True image-guided surgical planning. Enables the surgeon to precisely program the size, shape and location of each incision.

Indication

The LenSx® Laser is indicated for use in patients undergoing cataract surgery for removal of the crystalline lens. Intended uses in cataract surgery include:

anterior capsulotomy,

phacofragmentation,

creation of single plane and multi-plane arc cuts/incisions in the cornea, each of which may be performed either individually or consecutively during the same procedure.

Caution

United States Federal Law restricts this device to sale and use by or on the order of a physician or licensed eye care practitioner. United States Federal Law restricts the use of this device to practitioners who have been trained in the operation of this device.

MACULAR DEGENERATION

The macula lutea is a 5 mms yellow spot near the center of the retina at the back of the eyeball. It is at the end of the *visual axis*, the place where the concentration of vision is greatest, so the eye is in constant motion from side to side to bring a greater proportion of the visual field to the macula. Here there is found the greatest concentration of sensitive elements for recognition of light and transmission of nerve impulses and half of the brain's visual cortex is dedicated to its messages.

Age-related Macular Degeneration (AMR)

AMR is a major cause of loss of vision in the elderly, the term *senile* has been used to define it, bur is now considered pejorative and has been discarded. There are dry (*atrophic*) 90%, and wet (*exudative*) 10%, forms. In the dry form there is a deposition of yellow pigment (*drusen*), of uncertain origin, significance and nature; they are probably not the cause of the problem, merely a concomitant.

In the less common wet form there is hemorrhage and fluid, resulting from an abnormal growth of blood vessels deep to the retina, lifting it and possibly causing a retinal detachment. The patient may not notice any abnormality or may recognise a blank spot in the center of his visual field; more probably the condition is diagnosed at routine eye examination with the ophthalmoscope when pigment (or blood) is seen deposited at and around the macula.

It is found in 10% of patients 66 to 74 years of age, increasing by 30% in patients 75 to 85 years of age; it occurs with greater frequency in those with Caucasian heritage compared with Africans, more frequently in smokers, there is a heritable tendency such that a person with a close relative with AMD has a 50% chance of developing AMD, compared with the average 12%; it is associated with a high fat intake; beyond that little is known about its cause.

Although the loss of vision may take the person to below 20/200, and make the technically and legally "blind," it does not progress to complete loss of vision.

Pathology and Treatment

In the dry form of AMD there is found an atrophy of the pigment layer below the retina, resulting in a loss of both rods and cones, and a general diminution of central vision. No treatment has yet been established as of value, which does not mean there is no treatment offered – vitamins, omega 3 fatty acids, stains, and antioxidants are popular for every condition that has no accepted remedy. The 2007 Cochrane Review of publications found they conferred no benefits on the patients suffering from AMD.

In the totally different wet form of AMD, a network of capillary vessels has developed (*choroidal neovascularisation*), and the resulting hemorrhage and scarring damage the neuro-receptors

Diagnosis and treatments of wet AMD

The condition may be obvious at fundoscopy (examination of the back of the eye with the ophthalmoscope) or may require the intravenous injection of a fluoroscein dye to make the abnormal vessels visible beneath the retina.

Direct treatment is by injection of drugs to counteract the abnormal presence of chemicals inducing blood vessel formation (vascular endothelial growth factor VEGF) and thereby prevent blood vessel formation (*anti-angiogenetic*) using Avasin or Lucentis injected into the vitreous humor in the eyeball behind the lens, and repeated at intervals of one month. The cost is great, the benefits are less obvious. There are also treatments with laser, first identifying the abnormal vessels by intravenous injection of dye, then turning the laser beam on it.

GLAUCOMA

Just as there is a "pressure" inside blood vessels, so is there a pressure inside the eye, this is easily confirmed by squeezing (gently) on the eye and realising it does not collapse. The usual intra-ocular pressure (IOP) varies between 11 and 21 mms Hg, (blood pressure varies between 80 and 120 in the young adult). The eye may suffer damage if the pressure is abnormally high, the condition of glaucoma, the second commonest cause of blindness in the USA, but a high intra-ocular pressure does not necessarily cause damage, and one out of six patients with glaucoma have an intra-ocular pressure in the normal range, although it's too high for their eye.

Normal physiology
The eye is divided into two compartments by the lens and the ciliary body that attaches the lens to the wall of the eyeball. Anterior to the lens lies the disc of the iris, with its central gap, the pupil. Between the iris and the cornea the space is known as the anterior chamber, rather confusingly the slight space between iris and lens is the posterior chamber (it's easier to forget everything behind the lens to understand this!). Between the ciliary body that secures the lens, and the iris that lies in front of it, are fronds of *ciliary processes* which are highly vascular, and have a *trabecular network* surface area of about 6 sq.cms. in each eye. They secrete electrolytes into the anterior chamber, by osmosis that draws fluid into the anterior chamber and a circulation of this fluid is set up. There is a structure, known as the *canal of Schlemm*, but this is not a canal in the usual sense, it's a circular thin walled vein in the margin of the eye at the

angle (key word "angle") formed between iris and cornea. Normally the intra-ocular fluid is taken into this vein which has unusually large perforations in its wall, and then flows on into the general venous system. If for some reason the fluid cannot exit the eyeball, pressure builds up and the optic nerve is damaged, possibly by the effect of pressure on its blood vessels.

Pathology
The condition of glaucoma is divided according to its cause, either as open or closed "angle." In the "open angle" variety, there is no apparent blockage at the canal of Schlemm and associated structures, whereas in the "closed angle" variety an obstruction can be found.

Symptoms
By the time there is a recognised loss of vision, the condition is far advanced. The condition should be diagnosed by routine pressure measurements at the annual "well-eye" examination.

Risk factors of open angle glaucoma
Known risk factors are: old age; positive family history; African heritage; diabetes; hypertension; myopia; steroid use.

Diagnosis
Although a raised IOP does not necessarily cause glaucoma, we fortunately have two eyes, and if the IOP is higher in one than in the other, glaucoma is the diagnosis until proven otherwise. Vision that has been lost will not return. The objective is to diagnose before there is any loss.

SURGICAL TREATMENT
OPEN ANGLE GLAUCOMA

These are about 90% of the glaucoma cases in the USA. Surgery is indicated when other therapy with medication has failed or is considered inappropriate. The single objective of surgery is to reduce the intra-ocular pressure, this may be effected by increasing the outflow of fluid, or decreasing formation of fluid. There is much variation in technique and in instrumentation, but the majority of procedures fall broadly into the two categories of Trabeculectomy (aka filtration surgery) which fully opens the drainage area and Trabeculostomy which opens it partially but has fewer side effects.

Trabeculectomy is a century old practice and opens the full thickness of the drainage area, a flap is cut into the sclera (white tough coat of the eyeball) a *sclerostomy* and part of the iris may be removed, *iridectomy*. Following surgery half the patients are freed from medication, the remainder may or may not have better control of the IOP.

Trabeculoplasty is performed by laser, it consists of burning a hundred tiny holes in the drainage area. It is not a definitive procedure in that the patients still need to take medication and half may need the 15 minute procedure repeated in a couple of years or so, and there are some patients in whom the IOP actually increases, with ill-effects.

Tube shunts may at times be required, these are half inch plastic tubes which drain fluid from the anterior chamber, through the eyeball to the eye socket from which it is absorbed. The outflow of fluid is not controlled, IOP may become too low with damage to the eye's interior, or the tube may become blocked and ineffective.

Deep sclerectomy is a procedure in which a flap is made in the outer sclera, a piece removed from the inner sclera, and the anterior chamber is not invaded.

WHAT'S A DIOPTER?

Every child has used a magnifying glass to cause the rays of the sun to converge on a single point and burn a hole in paper. Most older adults have used spectacles, but there are few among us who can interpret the prescription written for the correct lens to be inserted into the spectacles frame.

We are all used to measurements, be it in feet or meters – the power of magnification is measured in diopters.

The curved surface of the lens "bends" the rays of light as they pass through it, a process called refraction. In this process if the lens is convex, as is the lens of the eye, the rays of light are caused by this bending to converge on a point at some distance from the lens, the *focal point*. The distance between lens and focal point is the *focal length*.

If a lens is constructed so that the light passing through it converges on a point 1 meter distant from the lens, a focal length of 1 meter, then the *refractive power* of that lens is plus one diopter (written as +1). If the focal length is half a meter the lens has a refractive power of 2 diopters, and at 10 cms focal length, the lens refractive power would be 10 diopters.

Eyes do not have concave (hollow) lenses, but in physical systems employing concave lenses, their power of diverging light is measured in the same manner but on a negative scale (written for instance as -1).

Refraction in the Eye

Because of the name it is given, it is natural to suppose that the lens in the eye is the beginning and end of the mechanism of focusing light, whereas in fact it is only one part of the whole mechanism.

The *refractive index* of a substance is a measure of how readily light passes through it, the arbitrary standard of 1.0 is light passing through air. It is the slowing of light rays through a curved surface that causes them to become focused. In the human eye, before the light reaches the retina on which an image is "cast," it passes in sequence through several substances, which respectively have refractive indices of 1.38 (cornea), 1.33 (aqueous humor), 1.40 (lens), 1.34 (vitreous humor), collectively giving the eye a refractive power of 59 diopters. What is not generally appreciated is the lens comprises only a third of this total, the cornea contributes more than the lens, but only the lens can change its shape to "accommodate" to change the focus.

PRESBYOPIA

THE AGING EYE EXPLAINED

It's an inescapable fact, that if you don't die you get older. Equally inescapable is the deterioration of the structure of the body as we age, shared with all living articles, animal or vegetable, and probably also mineral. The only question is the rate at which it happens, faster in the human than in the Galapagos turtle.

Presbyopia is the change in the eye's function that appears to cause the arm to lengthen, as we age– the object we read or examine has to be held further and further away from us. Put in technical language, the farsightedness (*hyperopia*) is related to the decreasing ability of the lens to change shape (*accommodate*) to focus on objects at varying distances from the eye.

Mechanism of accommodation

The lens progressively changes its physical structure through life. In the very young the lens has a strong elastic capsule filled with viscous proteinaceous fluid, and if removed from the eye becomes spherical, like a rubber ball that is no longer stretched. Within the eye the lens is held in place by as many as 70 suspensory ligaments (*Zonule of Zin – those eye people have fascinating names!*), running from the ciliary muscle to attach to the margins of the lens; tension transmitted through these ligaments gives the lens its more flattened shape, a shape altered as needed to focus on near or distant objects.

Aging changes in the lens

The nature of the protein in the lens changes with advancing age, the lens becomes larger, thicker, and less elastic. As a result the child's ability to accommodate, in the region of 14 diopters, has decreased in the 50 year old adult to 2 diopters, and in the 70 year old adult there is virtually no power of accommodation remaining – the state of presbyopia.

What to do?

The traditional compensation has been by wearing spectacles (glasses), first reading glasses when required, then bifocals or multifocals worn throughout the day. A more recent option is surgical correction.

PRESBYOPIA SURGERY

Correction for presbyopia may be undertaken surgically, the procedure performed either on the lens or the cornea, remembering that the cornea has in fact more refractive effect in the eye than has the lens.

The Lens:

Refractive Lens Exchange (RLE)

Not only does one lose the ability to accommodate as age advances, but the lens tends to cloud with the condition called *cataract*. If the surgery necessary to preserve vision is to be performed for cataract by an exchange of the normal lens with an artificial one, the surgeon may offer a variant which will assist also with the condition of presbyopia. In fact this type of lens might be exchanged before cataract has set in, employing an accommodating or multifocal intraocular lens.

The Cornea:

Monovision LASIK

It is normal to have a dominant eye, although both are working together to give stereoscopic vision. In the "monovision" procedure the surgeon, using the laser, reshapes the cornea of one eye to allow close vision while the other eye remains "farsighted." Some patients find their state confusing, some adapt to it well. It is possible to try lenses in the eye or spectacles prior to surgery to find whether one would be benefited.

SUPRACOR procedure

Performed using a high-tech laser system called the TECHNOLAS Excimer Workstation. This excimer laser is used to precisely reshape the cornea to restore near vision and simultaneously treat hyperopia or myopia, if necessary. The treatment involves first creating a thin flap on the surface of the eye with either a femtosecond laser or a microkeratome. This upper layer or flap is then moved to one side to allow the surgeon access to the cornea. Then the surgeon uses the excimer laser to accurately reshape the cornea and treat the presbyopia. Not yet available in the USA, but in use in Europe where it was developed.

Multifocal LASIK (PresbyLASIK)

As in an artificial lens, with the use of the laser, separate zones are created on the cornea, to provide for near, intermediate and far vision.

Conductive Keratoplasty (CK)

Conductive keratoplasty is a non-laser form of refractive eye surgery designed to help people reduce their need for reading glasses after they become presbyopic. It uses low energy radio waves to reshape the cornea and restore near vision. Radio frequency (RF) energy is directed to specific spots that form a circular pattern on the outer part of the cornea, the tissue shrinks, changes the curvature of the cornea, and brings near vision back into focus. This procedure might be performed on one eye only, creating a state of "monovision," near in one eye, far in the other.

The benefits of CK may not be lifelong since presbyopia is an advancing condition. It might, however, serve as a stop-gap delaying the need for reading glasses, and filling the interval before replacement lens surgery for cataract is indicated.

Corneal Inlays and Onlays

Corneal inlays and onlays are implantable devices that are surgically inserted just under the superficial layers of the cornea. Because the corneal implants alter the way light enters your eye, near vision is improved. They are at the clinical trial stage in the USA.

PTOSIS OF THE EYELID

Ptosis means a drooping, when applied to the eye it may indicate one or both eyelids are drooping.

Anatomy

The eyelid is relatively heavy, it has to be kept raised by an active muscular force, and when the muscle relaxes, as in drowsiness, the lid begins to droop. The central portion of the lid which gives it a firm consistency is a plate of dense fibrous tissue (*superior tarsal plate*) curved to the shape of the eyeball, covered in front with loose skin and lined behind with conjunctiva.

The *levator palpebrae superioris* is a flat sheet of muscle which raises the eyelid, it is attached to the bone of the upper orbit, and extends into a broad tendon attached to the tarsal plate. The *orbicularis oculi* muscle is a circular structure around the orbit, and this closes the eyelid tightly as in squinching up the eye. The 3rd cranial nerve (oculomotor) gives the eyelid muscle its nerve supply, nerve injury is one of the causes of a drooping lid, found also with injury to the sympathetic nerve supply (Horner's syndrome).

Clinical

Ptosis may be congenital (hard to explain), may be associated with specific injury locally, or with a generalised medical condition, or with drug abuse. Most often it is associated with advancing age, fatigue of the muscle and stretching of the tarsal plate that should hold up the lid.

Treatment

When the cause of the problem can be identified, treatment must first be directed to the cause. When treatment of the result of the problem, frequently age, is all that is left, it generally means some form of surgery, known as *blephoroplasty*. Non-surgical options exist, with the use of acids and lasers but are limited to the earliest of cases.

Incisions are made along the creases of the upper lid, excess fat is removed, the muscle and fascia may be tightened up, and excess skin folds are excised. Obviously a part of the skill and judgement rests in making the two eyes look the same.

THE DIABETIC EYE

Type two diabetes (T2DM) improperly called "late onset" diabetes, is increasingly revealed, and is generally associated with obesity. Unfortunately it is also associated with destructive changes in many parts of the body, and these changes do not make themselves known until they have reached a point beyond reversibility. The eyes are among the organs involved, and blindness is associated with diabetes; although caused by a number of different mechanisms, most are related to small blood vessels, the target area for diabetes induced pathology.

Diabetic Retinopathy

After 20 years almost every type 1 diabetic (improperly called juvenile diabetes) will show some degree of diabetic retinopathy (damage due to disease in the retina), as will half of type 2 diabetics; diabetic retinopathy is now the commonest cause of blindness in North America. Pregnant diabetics are particularly at risk, but sufferers from the chromosomal abnormality of Down's syndrome are exempt (for a reason yet to be discovered).

There are two forms: non-proliferative and proliferative (meaning sprouting new blood vessels). In the non-proliferative variety there are leaks of blood and fluid, or vessels close, but blindness usually does not occur. In the proliferative form the new vessels grow like kudzu, damage the retina, fill the vitreous, enter the front of the eye, scars form, the retina tears and vision is lost. It is treated by photocoagulation using a laser beam to coagulate the inappropriate blood vessels.

Macular Edema

Fluid leaks from the blood vessels obscuring the most sensitive part of the retina, resulting in blurring of the central field of vision.

Glaucoma

Diabetic retinopathy is a precursor to glaucoma, increased intra-optic pressure, which results in damage to the optic nerve before the patient is aware he has a problem.

Cataract

There is no question but that diabetics are prone to cataracts, both in greater numbers and at a younger age. Why that should be so, since the lens is one of the few parts of the body that does not have blood vessels, remains uncertain. It is probably related to a sugar alcohol (*sorbitol*) and an effect on fluid absorption into the lens.

Prevention of complications

The diabetic must keep his blood sugar under control, and he must have regular eye examinations by a skilled person who knows what to look for, and who can treat, or arrange treatment, of the complications observed at an early stage, long before they become symptomatic and irreversible.

PTERYGIUM

A growth that forms on the conjunctiva, at either side of the pupil, most often the inner (nasal) aspect. It may enlarge, invade and cause distortion of the cornea, changing its refractive nature or result in an astigmatism. It is found more frequently in geographic areas of brilliant sunshine as indicated by the term "surfer's eye." As the name implies, there is a body, "wings" and an advancing head.

Treatment

Those likely to be exposed to grit in bright light, typically on a sand beach, should wear protective goggles as a preventive measure.

A pterygium does not necessarily have to be removed, it is locally invasive but is not thought of as "cancer" and does not spread to other parts of the body. If removal is indicated by increasing size, local irritation or unsightliness, it is excised, no medication is available to deal with it..

Recurrence

There is a high recurrence rate, of up to 40%, because of this an anti-metabolite drug might be indicated for local application. To prevent recurrence after surgery a local application of radioactive strontium might be indicated. Auto-grafting with conjunctiva from the eyelid, or grafting with amniotic sac might be undertaken.

[Pterygium should not be confused with a Pinguecula, a yellow-white mass in the same position but which does not spread and does not distort the cornea.]

RETINA

FUNCTION and STRUCTURE

In very simple terms: the function of the retina is to transform the energy of the light rays which have entered the eye, into electrical impulses to be transmitted by the optic nerve to the brain.

The anatomists describe 10 distinct layers of cells in the retina, and it is easy to think of it as "inside-out" for the light sensitive cells are in the deepest layer, and the rays of light have to penetrate the superficial layers before they are recognized.

The layers of cells listed from the outer (deep) to inner (superficial) are:

1) Pigmented layer
2) Layer of rods and cones
3) Outer limiting membrane
4) Outer nuclear layer with cells for rods and cones
5) Outer plexiform layer
6) Inner nuclear layer
7) Inner plexiform layer
8) Ganglionic layer
9) Optic nerve fibers
10) Inner limiting membrane

One could easily be forgiven for thinking the retina recognized its own back to front disorganization. At the point where the light rays are most directly focused on the back of the eye, (a spot called the *macula*) for the width of a third of a millimetre, the superficial layers of the retina are pulled aside and only the most sensitive light receptor cones and a few rods are present; the rays of light do not have to penetrate the layers of cells here as they must elsewhere in the retina.

There's something about the expression, "rods and cones" that leaves those words in the minds of every student of biology. They are in fact wonderful structures. Each is a cell, and like most cells, each has a nucleus and cytoplasm with mitochondria, the "energy factories" of the cells, contained in the "inner segment" of the cell. What distinguishes them is the "outer (deeper) segment" of the cell, and the inner membrane of each folded into a hundred or more "discs" that carry the light sensitive photochemicals, and comprise 40% of the bulk of the outer segment portion of the cell.

The deepest layer bears an altogether different pigment, *melanin*, black in color and essential to prevent light from bouncing off the back of the eye; albinos who lack this pigment cannot tolerate bright light. Also stored in this layer is Vitamin A, essential for the formation of the photochemical, but unlike the wartime disinformation, consuming quantities of carrots does not make fighter pilots see better in the dark!

The more superficial layers bear intermediary nerve cells and their processes which carry the electro-chemical messages to the optic nerve and thence the brain.

The retina has two levels of blood supply. One of the reasons your doctor will use the ophthalmoscope to look into your eye is here is the only place in the body where it is possible to look directly at a blood vessel. The *retinal artery* is brought to the eye by the optic nerve, and where this enters the eye at the optic disc, the artery emerges, then gives off an upper and lower branch, each of which gives off a branch to the nasal (medial) and temporal (lateral) portions of the retina. These branches are *end arteries* meaning they do not join with each other, they have their own fields of retina to supply, and if blocked then that portion of the retina loses its blood supply.

The deeper part of the retina, bearing the rods and cones, gets its oxygen by diffusion from the blood vessels on the choroid, the layer of the eye on which the retina rests and to which it is attached.

RETINA — DISEASE

Retinoblastoma (cancer of the retina)

Cancer is never acceptable, but when it occurs in a young child it is particularly distressing. Although rare, retinoblastoma is nevertheless the most common of inherited childhood cancers. It comes in approximately equal versions – inheritable, and sporadic; it occurs usually in one eye, sometimes in both. Looking for signs of the tumor is part of a "well baby" examination, but they are not always easy to detect. As in all treatments, the order of importance is life, limb, function, appearance, and removal of the eye (*enucleation*) is often considered necessary.

Hypertensive Retinopathy

Associated with the various forms of abnormally raised blood pressure, essential and malignant hypertension, and the kind that comes with toxaemia of pregnancy. The walls of the retinal vessels thicken, the walls leak fluid or blood, "flames" of blood spots and exudates are visible with the ophthalmoscope.

Diabetes Mellitus

Diabetic retinopathy is a major cause of blindness related to both type one and two diabetes. In the "non-proliferative" form of diabetic retinopathy there are haemorrhages and exudates. In the "proliferative" form, blood vessels sprout and invade other parts of the eye.

Age Related Macular degeneration

A major cause of loss of useful vision, taking out the center of the visual field but leaving peripheral vision intact. There are described a "dry" form, and the less common "wet" form.

Retinitis Pigmentosa

This is an inherited condition which affects the rod cells of the retina, initially noticed to reduce night vision, then as it advances it is found central vision is reduced.

Retinal Blood Vessel Occlusion

Because the arteries are "end-arteries," blockage in the central vessel or one of its branches results in a sudden but painless, partial or complete, loss of vision in that eye. It typically results from a piece of the inner wall coming free (*embolus*) and blocking the artery downstream, as in a coronary occlusion, but there are other causes. Occlusion of a retinal vein is associated with disease processes such as hypertension or diabetes, and is less sudden in onset.

Retinal Detachment

A small or large section of the retina may separate from its attachment to the choroid, the occurrence is painless, but a segment of vision is lost and requires early treatment to ensure vitreous humor does not get under the detached portion and cause it to worsen by further detachment. There are a number of different types of procedures employed, but the underlying purpose is to reattach the retina to the choroid and to prevent further dehiscence.